DÉJÀ VU, ALL OVER AGAIN

DÉJÀ VU, ALL OVER AGAIN

The James Webb Space Telescope
and The Edge of Space:
Debunking the Big Bang

PETER STRASSBERG, M.D.

FCP

Full Court Press
Englewood Cliffs, New Jersey

Published in the United States of America
by Full Court Press, 601 Palisade Avenue,
Englewood Cliffs, NJ 07632
fullcourtpress.com

ISBN 978-1-953728-12-8
Library of Congress Control No. 2023900695

Editing and book design by Barry Sheinkopf

*Cover art, the Pillars of Creation, trunks of interstellar gas
found within the vast Eagle Nebula 6,500 lightyears away from
Earth, in which new stars are constantly emerging
(Webb Telescope image, courtesy NASA).*

TO MY FRIENDS,

Fear not the twilight years.

Table of Contents

Preface

The James Webb Space Telescope is a remarkable instrument. It greatly expands our ability to see objects both within and beyond our solar system. It will allow us to find other Earth-like planets orbiting distant stars. It will also enable us to peer deep into space, close to the very "beginning" of time.

Most scientists believe in the Big Bang theory—that the universe began in some giant explosion. If true, then remote galaxies (at almost the beginning of time) should appear relatively immature, and those farthest away should be found in the earliest stages of development.

However, astronomers will see something quite different. They will find that truly distant galaxies are the same as those close-by. Their cherished Big Bang theory, a concept glued together with multiple *ad hoc* abstractions, will finally unravel. In its place a simpler idea, a fourth-dimensional, hypersphere universe, will take hold, freeing science from a 100-year-old "straight jacket." We can all, then, begin our new quest—a journey toward a space-faring future.

Chapter 1

Introduction

THE JAMES WEBB SPACE TELESCOPE (JWST) marks the dawn of a great adventure. It will allow the exploration of the very edges of our universe, the earliest epochs of time.

Since most believe that the universe had a cataclysmic start—a Big Bang—some 14 billion years ago, by looking deep into space we should see the world's earliest galaxies, still in their infant forms.

But will we? I predict, instead, that we will merely find what is already seen close-by: galaxies and their clusters fully matured. We will simply see *more of the same*.

Why, then, are current beliefs so mistaken? Was there really a Big Bang? I will show that the whole concept is wrong and I will give a simpler, more logical version of what truly exists.

Einstein and Lemaître

The Big Bang concept started with Albert Einstein's theory of relativity. It was interpreted as showing a universe that was either expanding or contracting. Georges Lemaître, a Belgian priest and physicist, noted in the late 1920s that, if it was currently expanding, it had to have been smaller in the past. Therefore, at its earliest stage it would have been some tiny entity. He felt that this "primeval atom" had "burst" out to become today's world. There had been a cataclysmic beginning—in essence, a Big Bang.

Einstein, however, hewing to the accepted beliefs of his era, felt that the world had to be stable. To prevent the unwanted expansion or contraction inherent in his formulas, he inserted an arbitrary term—a cosmological constant—to maintain this stability. Although he accepted Lemaître's math, he balked at its implications.

Redshift

Later on, in the early 1930s, when Einstein journeyed to the United States, he met with the renowned astronomer Edwin Hubble. He was shown Hubble's findings of redshift changes in distant galaxies. Hubble noted that the farther away galaxies were, the more their light waves expanded—they shifted more toward the red spectrum (the widest wavelengths of visible light). Hubble felt that this represented a

Doppler shift, something readily noted in the changing pitch of a passing ambulance's siren. As the ambulance approaches, the pitch increases because sound waves contract. As it departs, the pitch lowers because sound waves expand.

In a similar fashion, light waves from a cosmic object would contract or expand depending on whether it was approaching or departing. Astronomers had already noted that stars coming in our direction shifted their light toward the blue (or shorter) wavelengths, and those moving from us toward the red (or longer) wavelengths. Also, the speed of approach (or departure) would exaggerate this shift: the faster, the greater the change.

Hubble felt that his findings (the farther away, the greater the shift) represented an expanding world. Once made aware of this data, Einstein realized that Lemaître's concept could be correct, and that his own cosmological constant had been an error. He later deemed it the "biggest blunder" of his career.

However, although Hubble's data allowed for expansion, the concept of a Big Bang did not at first catch on. Hubble had miscalculated the rate of expansion by a huge factor, estimating the age of the universe at about two billion years. Since the Earth was known to be over *four* billion years old, this was obviously wrong. So although expansion was assumed correct, the idea of an initial cataclysm was shunted aside.

Other ideas incorporating expansion were considered, the most important being the astrophysicist Fred Hoyle's steady-state theory. He felt that, although there was a constant expansion of space (filled by a continuous formation of new hydrogen atoms), there was no identifiable beginning. In fact, he disparaged the whole belief of an explosive start, dismissing Lemaître's theory as a "big bang," a name that would thereafter stick.

Cosmic Microwave Background

These competing beliefs were hotly contested until the mid-1960s when two scientists (at Bell Labs in New Jersey) using a radio telescope noted annoying static in their equipment they could not eliminate. They finally concluded that it came from cosmic radiation found equally in all parts of the sky.

Since this radiation was found in the microwave range (wavelengths of $1/1000^{th}$ of a meter) it became known as the cosmic microwave background (CMB).

Other theorists had claimed that, if there had been a Big Bang, the explosive start would have spewed out much smaller, infrared photons (wavelengths of $1/1,000,000^{th}$ of a meter) that, over the millennia, would have expanded with the rest of the universe at least 1,000 times to their current size. Hence, CMB radiation was proof of the Big Bang. Hoyle's stead-state theory could not account for this background and was discarded.

Inflation

The Big Bang became the accepted theory; however, there were still significant concerns. A "messy" beginning explosion should not be able to account for the surprising homogeneity of the current universe. Also, parts of the universe too distant from each other to have ever been in contact still showed the same background energy. Why would these distinct, distant areas be similar?

Alan Guth, an astrophysicist, came up in the late 1970s with an intriguing solution to these problems. He theorized that, very shortly after its initial cataclysmic start (within the first trillion, trillion, trillionth of a second), the universe rapidly inflated. This inflation evened out the initial messy start, just as a crinkly balloon becomes smooth once blown up. Areas currently too distant to have been in contact, however, had been, prior to inflation, much closer together and now presented with the same energetic background.

A reason for this inflation could not be given, but, since it so effectively answered those important doubts, it allowed the Big Bang to continue as the mainstream theory.

Dark Energy

Later, in the 1990s, another question arose. If gravity was the only cosmic force, could the universe's supposed expansion be ending? Could it possibly be contracting under the constant pull of gravity? To answer this concern, accurate distant measurements were needed. A special kind of supernova (Type 1a) was used to now determine the universe's true size and see if it correlated with Big Bang theory.

These Type 1a supernovas are extremely bright. They could be seen over 10 billion lightyears away. They also all shone with the same intensity. They therefore could be used as "standard candles," their dimness revealing their true distance away.

When these were discovered and evaluated, they proved the opposite of what was thought. The supernovas were found farther away than theorized. The universe was not contracting under the constant tug of gravity. Instead, it was expanding and at an even greater speed than supposed by Big Bang theory. However, this added push or expansion required another energy source. Since none was known, that mysterious, unidentified source was named "dark" energy.

Today, the Big Bang theory is the accepted paradigm. Its basis is Einstein's theory of relativity and the belief in an expanding cosmos. Lemaître assumed an initial event, Hubble showed increasing redshifts, CMB buttressed it with a bath of microwaves, Guth smoothed it out with inflation, and dark energy negated gravitation's pull. The universe was not only expanding but its rate was actually increasing.

This is the current belief. . .but is it correct? The James Webb Space Telescope, which will allow us to look almost to the very edge of space, will give us a glimpse of some of the "earliest" events. It is believed that it will show us what occurred in an infant world: galaxies in their initial, most immature forms.

Nevertheless, I am stating the opposite. The JWST will show merely what we already see nearby—mature galaxies formed into giant clusters. There will be no "early" world; there will simply be *more of the same.* It will be, to quote Yogi Berra, "*Déjà vu,* all over again."

Chapter 2

Particulate Universe

Our world *must* have a smallest, finite size. This was first shown "inadvertently" by the Greek philosopher, Zeno, some 2,500 years ago. He was part of a school of thought that considered motion an illusion. The world, to their thinking, was stationary.

He is best remembered for a famous paradox concerning the impossibility of movement. Succinctly, before one could go any distance, let us say 100 yards, that person had to travel one-half of the way, or 50 yards. But before going 50 yards, that same individual had to travel one-half again, or 25 yards—and so, on and on indefinitely. This would never end; thus, before moving, one needed to travel an infinite number of ever-smaller steps. Since it is impossible to move an infinite number of times, no motion was possible.

However, although this paradox appears logically correct,

we *know* that it is false. There is obvious motion in our world. In fact, all things are in flux; nothing is truly static. *Thus, the underlying assumption must be wrong.* There cannot be infinite divisibility. There *has* to be a smallest, finite size.

The real proof is that, if this were not so, then space would consist of zero-dimensional points. However, if that were true, no matter how many such points were placed side-to-side, there would *still* be zero dimensions. (Zero multiplied by any number, including infinity, remains zero.) Hence, if there were no smallest size, nothing would exist.

Planck Scale

The renowned physicist Max Planck, in 1900, developed the mathematical basis for the concept of smallest, finite size. In his honor, it has been named the Planck length: 1.6×10^{-35} meters. Since the speed of light is the fastest possible speed, the least time that can exist would be the time it took a light wave to traverse one Planck length: 5.4×10^{-44} seconds. Finally, as we conceive of the world as three-dimensional, the tiniest volume of space would be a sphere with a diameter of one Planck length. However, Planck showed that, although it would be the smallest possible part of space, it would have tremendous mass. Its mass would be almost 10^{20} (100 million, trillion) times that of the much larger, but still miniscule, proton.

Fourth Direction

If the Big Bang is wrong, what is right? What can take its place? How are we to explain Hubble's redshift findings? What is the true reason for CMB?

To find the answer we must change our perspective.

We all sense the world as three-dimensional. We know of but three directions—front to back, side to side, and up and down.

But what if there was a fourth direction? If so, then all the findings could make sense and no Big Bang would be required. There would be no "beginning," just a curve toward that higher dimension that could explain all.

Understanding the Fourth Direction

Now, if there is a fourth direction, how could we actually visualize it? We know that, as we add a dimension, let us say from one, or a straight line, to two, or a flat plane, we increase space by an infinite amount but in a direction 90^0 to the original construct. Thus, a one-dimensional line:

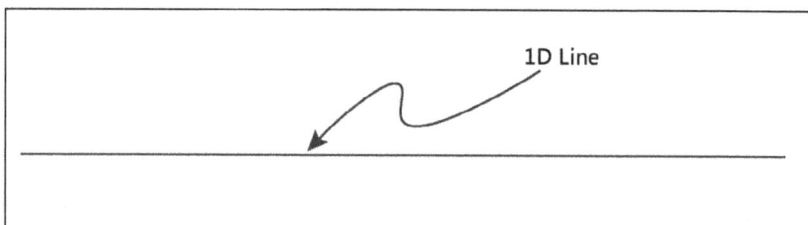

1D Line

becomes a two-dimensional plane by adding an infinite space at 90^0:

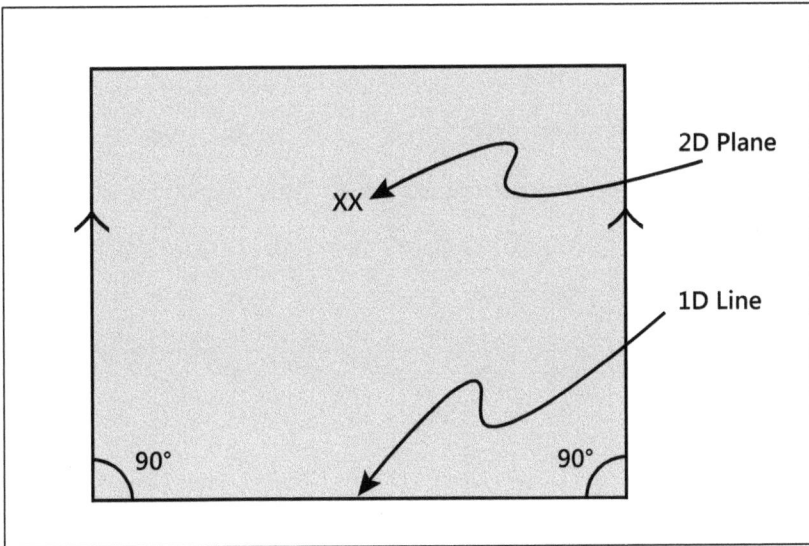

Similarly, a two-dimensional plane becomes a three-dimensional box in a like manner:

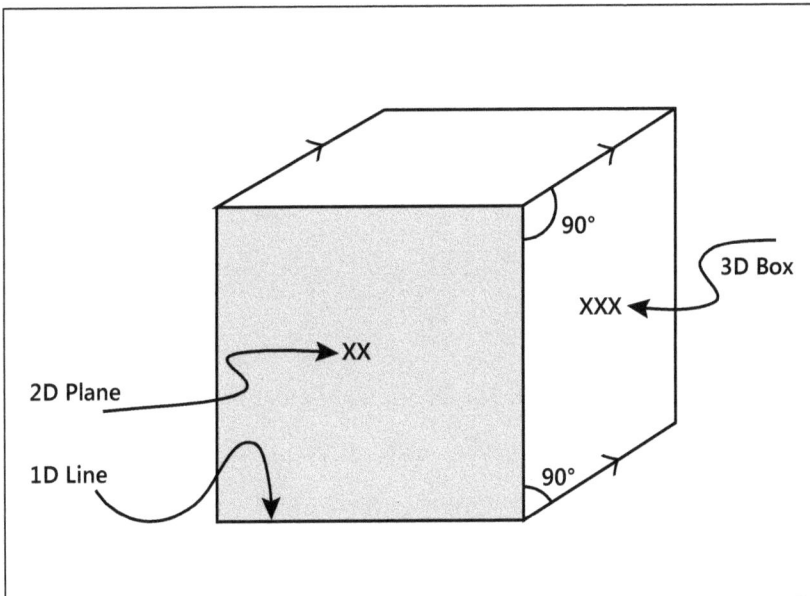

The fourth direction, therefore, must also be an infinite space but at 90^0 to the other three. Since we, being three-dimensional, cannot conceive of such a space, the fourth direction is, to us, imaginary. It becomes a potentially infinite space within each point that makes up our world. Each Planck volume, therefore, has another direction that extends inwardly some immense distance. Or, to draw it we get:

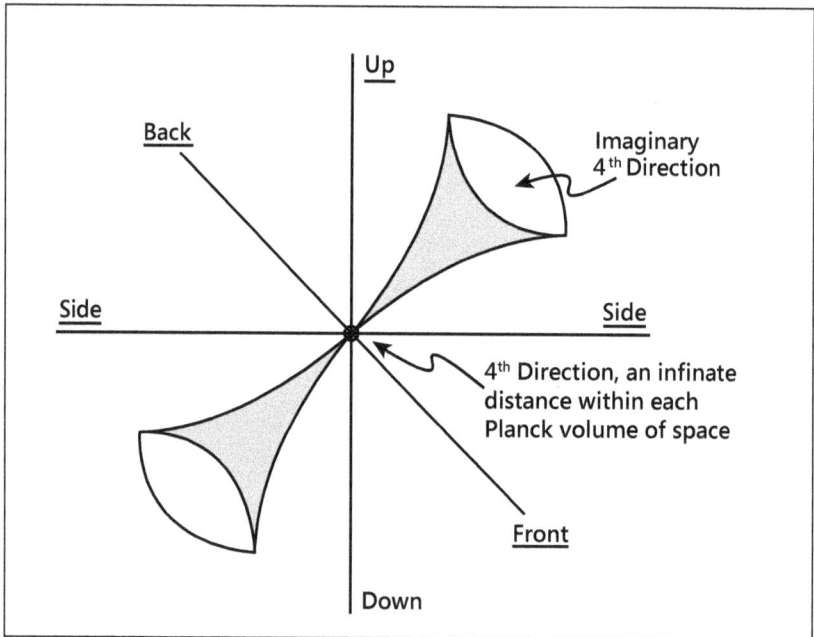

Finally, since each Planck volume or point of space is extremely massive, could this vast internal distance represent that enormous mass? Logically it would seem so. Since Einstein showed that mass and energy are the same ($E=mc^2$), then the true energy of the world is the higher dimension encapsulated in each and every Planck volume.

Chapter 3

Spheres

WE ARE THREE-DIMENSIONAL. Because this is so, and because everything that we see around us is three-dimensional as well, we have great difficulty in visualizing another, or higher, dimension beyond our own.

However, that dimension can be represented mathematically by using spherical objects. In math what we commonly consider a ball (a baseball, for example) is actually defined as a "2-sphere" (two-dimensional surface) covering a "3-ball" (three-dimensional interior). The surfaces of all round objects are called *spheres*, numbered by the dimensions they represent; their centers are named *balls*, always one dimension greater.

A straight line is one-dimensional. However, if we draw a circle, we still have a one-dimensional line but now curved about a higher dimension. It becomes a 1-sphere surrounding a 2-ball:

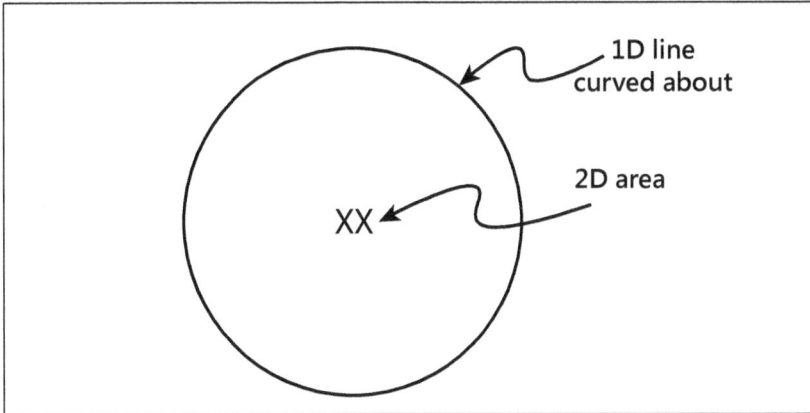

Likewise, a common, everyday ball or globe really describes an object with a two-dimensional surface curved around a three-dimensional volume. Thus, in mathematical terms, we have (as already noted for the baseball) a 2-sphere (two-dimensional exterior) surrounding a 3-ball (three-dimensional interior):

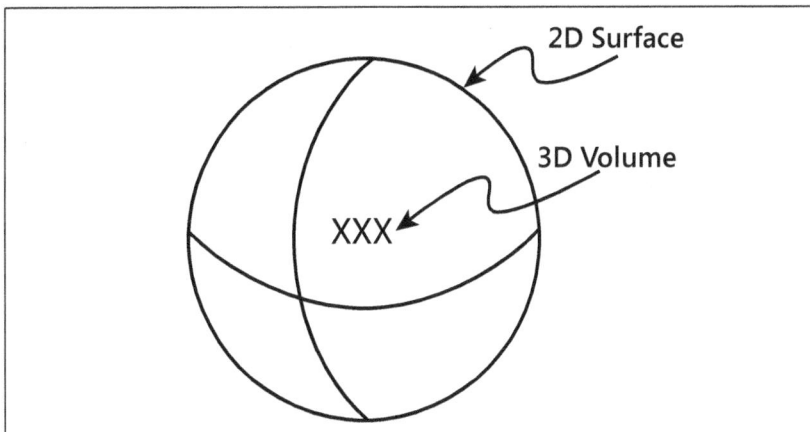

If the universe is actually four-dimensional, then what we consider our *three*-dimensional world is really a surface encapsulating a *four*-dimensional interior. It is a 3-sphere covering a 4-ball. We, and everything that exists, are merely the exteriors of a higher-dimensional core and, just as the straight line and flat plane were "curved," our world is also "curved." We simply have great difficulty comprehending the direction.

Lower-Dimensional Model

The Big Bang theory came directly from Einstein's general theory of relativity. However, it did not gain any traction until the redshift data was interpreted by Edwin Hubble. He felt, naturally, that the redshift increase was due to a Doppler effect, as that had already been observed in the light from nearby stars. He concluded, therefore, that distant galaxies were separating from us and each other, and that the greater the distance the more rapid was this separation. The universe was flying apart: It was expanding.

If, instead, we adopt a higher-dimensional curve, we can see exactly the same redshift change. Consequently, there is no need for any expansion, hence no requirement for an initial cataclysmic start or Big Bang.

However, as it is quite difficult, if not impossible, to visualize a higher dimension, we will use a lower-dimensional model to aid our understanding.

Let us assume that the world is one-dimensional. It, then, would be a simple straight line populated by segments of that line. But if it is really curved into a circle (1-sphere) then its "inhabitants" begin to see unusual changes. The farther out they look, the more stretched objects in their world appear. If they observe light from a distant source, its wavelength will be expanded. It will be redshifted. Let me draw this:

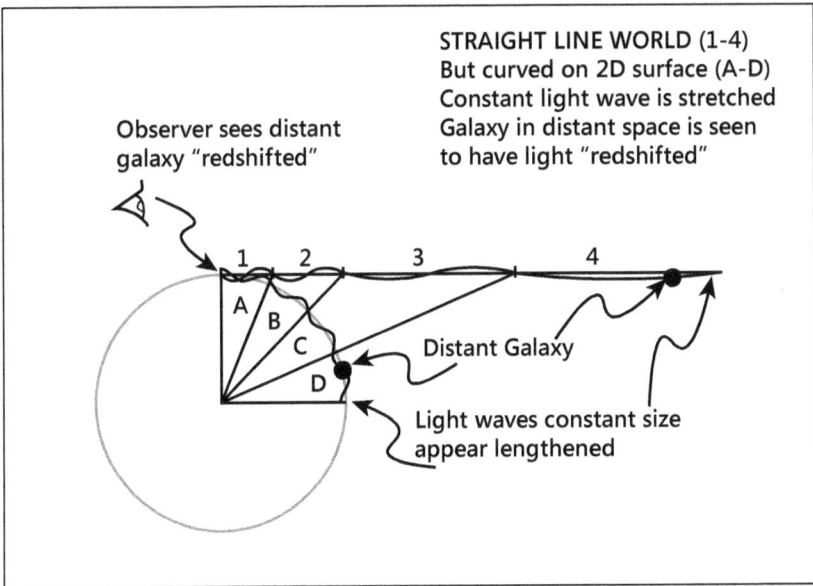

We see the same thing in a map of the Earth (Mercator projection). The farther north or south, the greater the distortion or stretch. In fact, when we get to Antarctica, it extends the entire width of the map. In this example, a higher dimension has been drawn as a lower one (three dimensions as two). This contraction of a dimension leads to an inherent distortion.

James Webb Space Telescope (JWST)

The redshift, then, is really just the same change. As we gaze ever farther away, the curve becomes more and more pronounced. Light expands from its normal wavelength into the red, then infrared, spectrums. The JWST will give astronomers an ability to clearly view objects over 13 billion light-years away. It is estimated that it will allow us to discern things up to 400 million lightyears from the supposed start of the universe.

What most think will be found are early, immature galaxies; formations just starting to evolve. However, that is wrong. Instead, they will find what is seen nearby. There will simply be *more of the same.*

It will take astronomers a while to fully appreciate this; for, initially, they will try to incorporate what they consider "unusual findings" into standard Big Bang theory. However, as that theory becomes ever more distorted, there will finally be a major paradigm shift.

Einstein had alluded to this when he showed that gravity could be understood as a bending or warping of space. He used the example of a two-dimensional surface—similar to that of the Earth's—to help in visualizing this distortion. He did so because of the inherent difficulty of picturing a three-dimensional bent surface. Further, he stated that the universe could curve back onto itself. Thus, theoretically, a spacefarer could travel out always in one direction, never turning, yet arrive at the same spot. He used the analogy of circumnavigating the Earth, heading east yet returning from the west.

Therefore, a higher dimension permits us to simplify redshift data; the example of a lower dimension allows us to better understand it.

Chapter 4

Necessity for a Higher Dimension

OES SUCH A HIGHER DIMENSION seem realistic? We know that our world is three-dimensional. Yet from a purely logical standpoint, any dimensional universe would have to exist in a higher-dimensional space.

A point, or zero-dimensional abstraction, needs to be a part of a one-dimensional line:

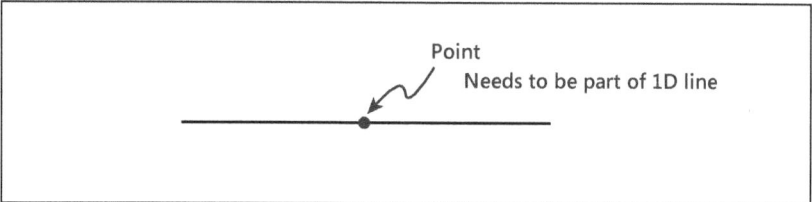

Point
Needs to be part of 1D line

A line can only be understood as drawn on a *two*-dimensional surface:

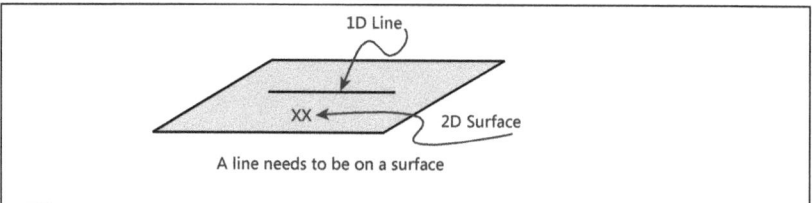

1D Line
XX
2D Surface
A line needs to be on a surface

A two-dimensional surface must be found within a three-dimensional space:

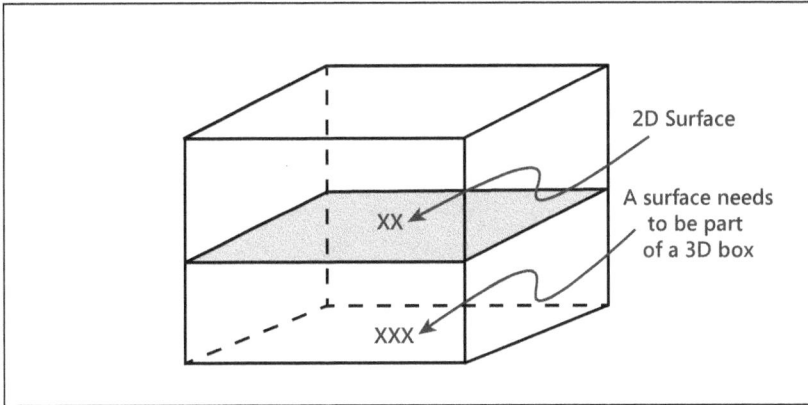

Likewise, a three-dimensional object requires a four-dimensional setting.

Drawing a Four-Dimensional Hypersphere

But how can we visualize this higher dimension? One way is to take a three-dimensional cube, flatten it (still keeping it three-dimensional), and then bend it "supposedly" toward the fourth dimension (*see first illustration on the next page*). We then place it on the surface of an unknowable four-dimensional "sphere."

Another way to try and visualize it is by slicing through a higher-dimensional sphere.

What is found is a lower-dimensional abstraction. Thus, if we slice through a globe (2-sphere) we get a circle (1-sphere) (see second illustration).

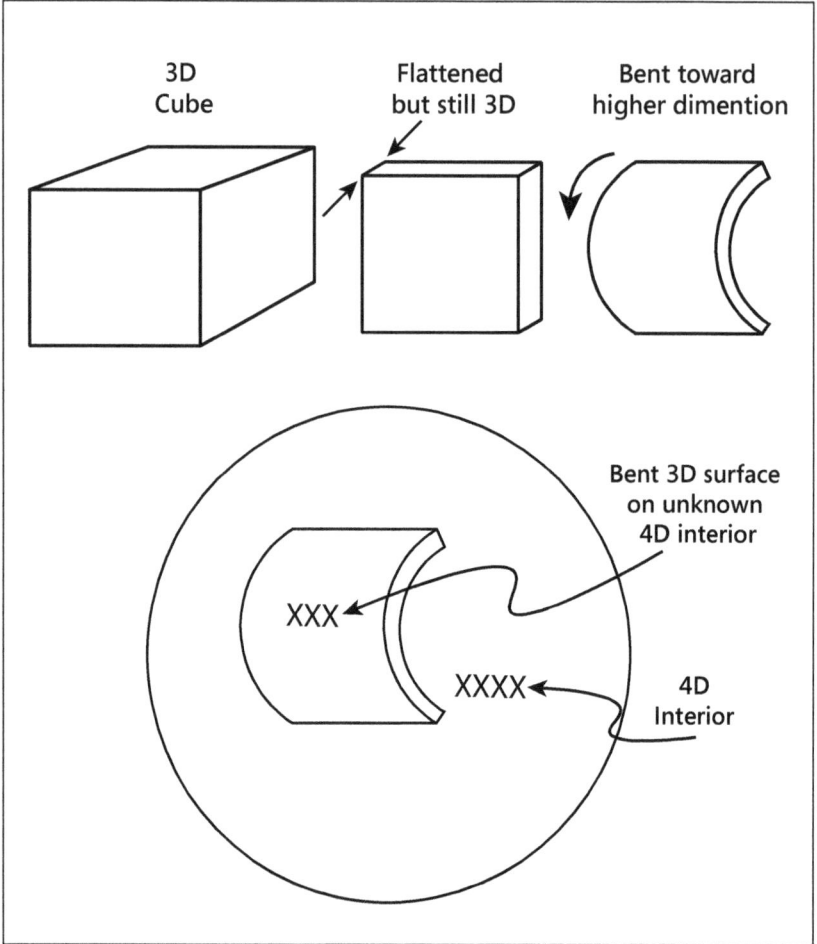

3D
Cube

Flattened
but still 3D

Bent toward
higher dimention

Bent 3D surface
on unknown
4D interior

XXX

XXXX

4D
Interior

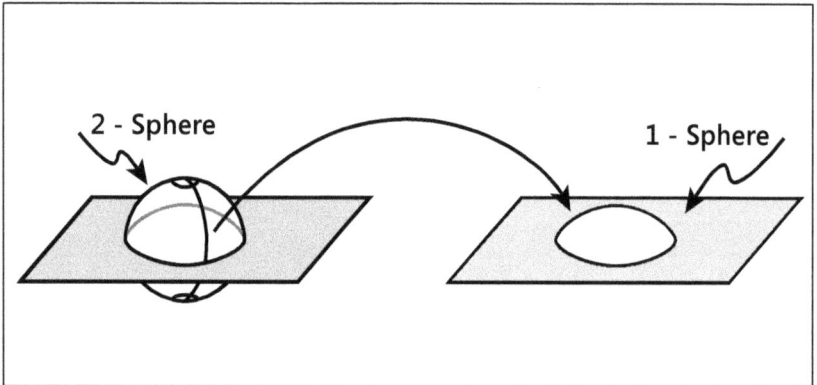

2 - Sphere

1 - Sphere

Hence, if we know what the lower-dimensional object looks like, we can extrapolate backward to the higher-dimensional entity. Therefore, just as a 2-sphere (globe) yields a 1-sphere (circle), a 3-sphere (hypersphere) gives us a 2-sphere (globe). So if we transect a 3-sphere, although we cannot *visualize* that 3-sphere, we get a completely well understood round object—a globe. We can then guess at what that hypersphere looked like in order to yield a globe. In this way we can begin to get an idea of a higher-dimensional object's appearance. Please see the following:

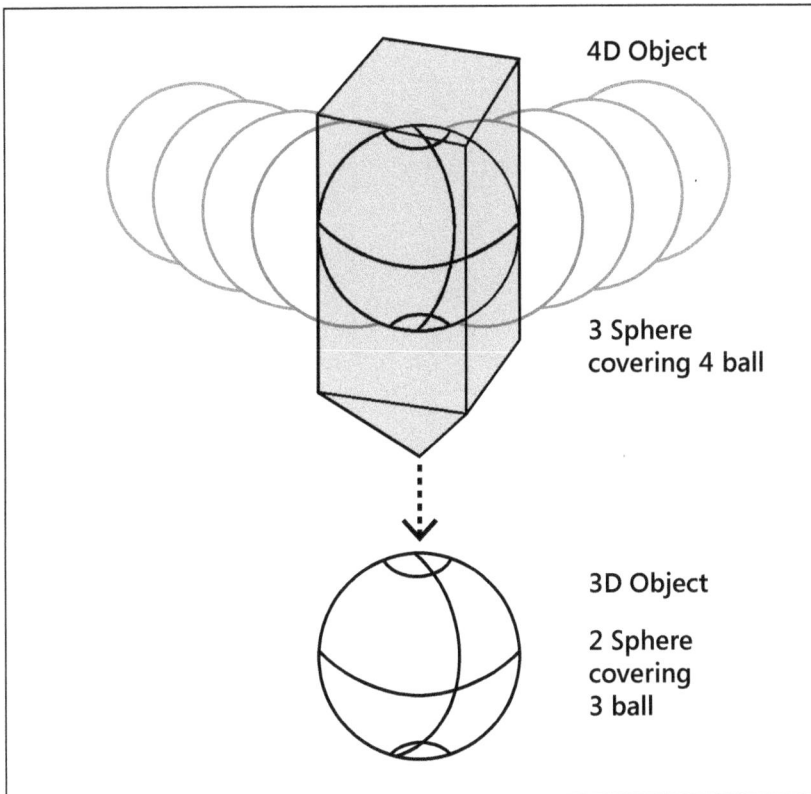

4D Object

3 Sphere covering 4 ball

3D Object

2 Sphere covering 3 ball

Einstein further noted that we must not think of space and time as separate entities. Instead, there is only "spacetime." Thus, the higher dimension, the dimension that is beyond the three dimensions of space, is time; and, if we slice spacetime we get a single moment of time—a distinct three-dimensional construct.

Chapter 5

Cosmic Microwave Background—
Another Explanation

THE BIG BANG THEORY WAS BASED on Einstein's theories; however, it only became mainstream after the Cosmic Microwave Background (CMB) discovery. CMB fit the criteria for a cataclysmic, expanding universe.

Nevertheless, could CMB be explained in another manner?

If the universe is a three-dimensional surface enclosing a four-dimensional core (3-sphere or hypersphere), the farther out we look, the closer we are to 90^0 on the curve. Objects appear to be continually stretched until 90^0; then, after that, all things disappear. No light can return from beyond that point. It is the "edge" of the universe or, as currently under-stood, the "beginning" of time.

Astronomers use the term "z" to connote this stretch. The greater the z, the wider the effect. At the farthest possible part of the universe, at just one Planck length less than 90^0, the z, or stretch effect, is enormous. It can be shown to be 10^{32} (or 100 million, trillion, trillion) times normal.[*]

We know that nothing can be smaller than a Planck length (10^{-35} m). If we were to stretch the Planck length by this enormous z, we would get microwaves—1/1,000th of a meter (10^{-35} m x 10^{32} = 10^{-3} m). Therefore, CMB is simply how we see the very edge of the world. The last visible thing must be microwave sized. That is why it is equal in *all* directions. Wherever we look, we see the same edge and the same stretch.

Furthermore, we can also explain why there are anisotropies or inconsistencies of 1/100,000th in CMB findings. What these imperfections consist of are the galaxies at this enormous distance, stretched by the distortion of the curve. They are not, as current theory suggests, quantum fluctuations

---

[*] *10^{32} is the stretch factor (z) of a light wave after which that wave would be greater than the size of the universe and therefore invisible. It is based on an average visible light wave of about 500 x 10^{-9} m or (5 x 10^{-7} m) and, when compared to the radius of the universe (10^{26} m) is found to be 2 x 10^{32} (10^{26} / 5 x 10^{-7} m) or, if rounded down, 10^{32} times smaller. Thus, a light wave is 1/10^{32} the size of the universe and, if stretched slightly more than 10^{32} times, becomes bigger than the world; it disappears.*

inherent to the Big Bang, which subsequently evolved into galaxies.*

Dark Energy-–Another Explanation

So far, we have been able to simplify the redshift and CMB data with the use of but one parameter—curved three-dimensional space. However, can we also clarify dark energy?

Dark energy was advanced to explain astronomers' findings of greater distances (to supernovas) than expected by Big Bang theory. However, once dark energy was accepted, it also allowed for a belief in the "flatness" of space.

Before the "discovery" of dark energy, scientists felt that there was insufficient mass to account for what, they believed, was a flat, unbending three-dimensional world. Visible mass—galaxies—could account for less than $1/20^{th}$ of the mass needed. Five times more was believed to be tied up in

---

*An average galaxy is about 10^5 (100,000) lightyears in size, whereas the universe is approximately 10^{10} (13,800,000,000) lightyears from any spot to its edge. Thus, a galaxy is around $1/100,000^{th}$ the size of the universe ($10^5/10^{10}$). Hence, galactic structures (essentially, swirling clouds of hydrogen atoms congealed into stars) in otherwise uniform space give rise to these irregularities. They make up $1/100,000^{th}$ of the entire volume of the universe; therefore, they present as irregularities of 1 part in 100,000. Current theory has this backwards.

an unknown quantity, "dark matter" (thought to exist in order to explain the rapid rate of rotation of galaxies and their clusters). However, even *with* dark matter, the mass required was still less than 1/3rd of what was needed. Nonetheless, once dark energy was postulated, the total energy became great enough to reach necessary mass. (Mass and energy are the same, $E=mc^2$.)

Thus, dark energy, at the same time, gave theoreticians both a cause for the supernova data and cemented a belief in a flat universe.

However, using a three-dimensional curve, one can easily explain the increased distances to these supernovas. They really are where they should be, given such a curving, three-dimensional surface. The table and graph on the following page explain this.

In the table, the distances calculated by using a fourth-dimensional concept (column 3) are greater and more accurate than those calculated through standard Big Bang theory (column 2) for all stretch or z factors (column 1) until just less than 12 billion lightyears. The difference in distance is what has erroneously been considered "dark energy." Please see appendix for details.

The same information is presented as a graph; the supposed dark energy becomes the area between the two curves.

COMPARATIVE DISTANCE
CALCULATIONS

1	2	3	4
Stretch Factor	Hubble/Expansion	4D Curve/Tangent	Dark Energy
("Z")	Dist. (10^9 Lys)	Dist. (10^9 Lys)	Dist. (10^9 Lys)
0.1	1.5	4.6	+3.1
0.2	2.8	6.1	+3.3
0.4	4.6	8.0	+3.4
0.5	5.5	8.5	+3.0
0.8	7.7	9.7	+2.0
1.2	9.5	10.7	+1.2
2.0	11.5	11.6	+0.1
2.4	12.3	11.9	-0.4

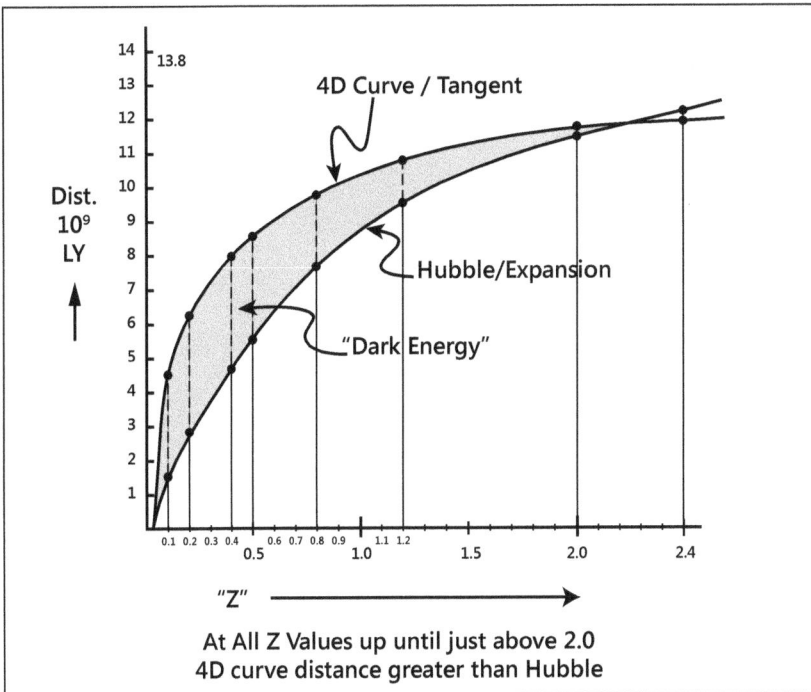

At All Z Values up until just above 2.0
4D curve distance greater than Hubble

Final Nail In The Coffin

What we have, now, is a more straightforward path. It gives a rational basis for the redshift and CMB data. It does away with dark energy and inflation. "Simpler" should mean more likely—"Occam's razor." However, simpler does not prove *correctness*. To do so, we will need Einstein's special theory of relativity.

According to that theory, velocity causes a contraction in size and an expansion in time. Thus, a speeding object's length will shorten, but its time intervals will lengthen. It will have longer seconds, minutes, or any time measurement when compared to us.

Now, these changes, although always occurring, are only easily observable when the velocity is extreme. Thus, if the object travels at a relativistic velocity (close to the speed of light—300,000 km/sec) these changes can be measured. (For example, if something moves from us at 85% percent of that speed [about 250,000 km/sec], its length halves and its moments double.)

Given the Big Bang theory, distant galaxies are supposedly moving away at such a velocity or more. Since the Type 1a supernovas are, not only extremely luminescent, but also of constant duration, they can be used to measure both distance and velocity. We know that these supernova events take about 20 days (start to finish). Consequently, if seen at a great distance, they should be moving away at an extreme, relativistic velocity. Thus, their times should have expanded; they should last at least twice as long (40 days). However, when measured, we find that they begin and end in the same 20-day period, no matter where observed. Thus, they exhibit *no* time dilation.

But, if there is no time dilation, then either Einstein's special theory of relativity is wrong (unlikely, as it has been proven correct in *all* situations) or these entities are not moving at a relativistic velocity. But if not so traveling, then the redshift is not a Doppler shift, and, consequently, the entire Big Bang theory makes no sense.

Hence, we have the final proof that the Big Bang theory is wrong. A simpler concept—with but one variable, a curved, three-dimensional world—now has been proved.

Chapter 6

Stars—Spheres
In Three Dimensions

S TARS, OUR SUN INCLUDED, EXIST as great hydrogen furnaces. Their hydrogen atoms are fused to helium, releasing vast energy. They shine with life-giving brilliance.

As the hydrogen, over billions of years, is slowly depleted, gravity contracts them into ever-tighter spheres. Finally, the thermonuclear reactions become too feeble to counter the overwhelming gravitational attraction. A star, then, can collapse with a spectacular explosion—as a supernova.

Stars, therefore, exist in a constant struggle of contracting gravitational versus expanding thermonuclear force.

Universe—Sphere In Four Dimensions

The universe is a hypersphere. It is curved, but in a way that is difficult to visualize. It, however, like its trillions upon trillions of stars, has a similar underlying dynamic. It retains its shape, not by the countering of gravitational attraction with thermonuclear energy, but, instead, with energized plasma.

Most matter in the universe consists of hydrogen (it makes up 90 percent of all atoms) either in gaseous form or as plasma.* The plasma found in deep space is quite hot and energetic. (Its average temperature is about one billion degrees.) Because of this energy, it exerts a "steam-like," expansive counterforce to the gravitational contraction of the stars. The final result, just as in individual stars, is a spherical shape.

This spherical shape, however, is unlike that of a common three-dimensional globe; there is no single center. Each proton is its center. It is a hypersphere, curved toward the fourth dimension—time. Hence, from any point, one can look outward (or backward) in time to the same edge—13.8 billion years ago. But that distance does not mark a beginning; it is simply the edge of the curve. Beyond it nothing exists. It is

---

*Plasma is the fourth state of matter. When an element is cold enough, it is solid. As it heats up, it liquefies, and then, when sufficiently hot, it morphs into a gas. Finally, when highly energized, the proton–electron bond is so loosened that it becomes plasma. At this point, individual protons are no longer in direct contact with their surrounding electron clouds.

always the same, a constant 13.8 billion lightyears away.

We find, then, in both individual stars and the universe, an inexorable tension between gravity's attraction and either thermonuclear or plasma expansion. Gravity is fundamental; a more thorough understanding, therefore, is needed.

Newton, Einstein, and Tesla

Gravity was described by Newton as an instantaneous, attractive force intrinsic to all forms of matter. Everything has this quality—gravitation—and it extends outward from all objects to all points in the universe. Nothing interferes with it. He also understood, however, that force required direct contact. How, then, could gravity be felt through the vastness of "empty" space?

Einstein described gravity as a warping or bending of space caused by massive objects. This was shown in the 1919 solar eclipse, where light waves (always traveling the shortest, "straight-line" distance) were bent by the huge mass of our Sun. He considered gravity to be simply acceleration, not an actual force. He equated the gravity we feel on Earth with what would be felt in a spaceship accelerating through empty space. Occupants of that vessel, when moving at a steady speed, would be "freely floating." Once the vessel began accelerating, they would immediately be pulled in the opposite direction – they would "fall to the floor." They could not differentiate this sensation or pull from the force of gravity.

Tesla, Einstein's contemporary and intellectual equal, felt that the concept of "bent" space made no sense. It was foolish, according to Tesla, to believe that something could act on nothing. He could not fathom how one could "bend" what was not there. These three great men of science all believed that there had to be "something" through which gravity acted. Empty space, they felt, could not really be "empty."*

We have already shown that the universe has a smallest size; there cannot be infinitely divisible space. The smallest object is the Planck volume. We also noted that something

---

*Although most scientists agree that Newton and Tesla believed in "something" inherent to empty space (allowing for the force of gravity), many feel that Einstein did not. In reality, we know that Einstein also strongly felt that there was something permeating space. In a 1920 speech, entitled "Aether in the Theory of Relativity," he stated: "Recapitulating, we may say that, according to the general theory of relativity, space is endowed with physical qualities; in this sense, therefore, there exists an Aether. [In fact,] according to the general theory of relativity, space without Aether is unthinkable; for in such space there not only would be no propagation of light, but also [no] possibility of. . .any space-time intervals in the physical sense" (emphasis added). (See Manning and Manewich (2019), Hidden Energy, Victoria, BC: Friesen-Press, p. 95.)

of this size has an enormous mass. The Planck volume's mass is 100 million, trillion times as great as that of the much larger (but still miniscule) proton. Yet, although empty space is composed of tiny, massive parts, this mass is not felt. We do not "bump" into "emptiness."

Newton's Laws

In order to begin to understand how gravity works, one must first employ Newton's basic concepts. His laws, in essence, state that an object at rest or in constant motion remains so, unless directly acted on by an outside force. He equates this force with acceleration. Thus, unless there is an acceleration (or a deceleration) of movement, no force is involved.

Newton further stated that every force is countered by an equivalent, opposing force: *every action has an equal and opposite reaction.* This is Newton's most fundamental law; for, without it, our universe could not exist. (Every push on something, a wall, for example, must elicit an equal push back. If not, a solid wall would not be "solid," and the universe would not be as it is.)

Inertia

Inertia, also described by Newton, is what is felt when force is applied. (When standing in a motionless subway car, we feel a sudden jerk as it starts to move. When sitting in an automobile, we are pushed back into our seat as it accelerates. These are forms of inertia.)

Newton did not try to explain the basis of inertia, he just felt that it was intrinsic to mass. However, mass is only felt (or exists) when there is acceleration. Therefore, since acceleration is essential for mass, it must also be essential for inertia. *Acceleration, then, gives substance and meaning to objects in the world.*

Hydrogen

The universe consists mainly of hydrogen. It makes up about 90 percent of all individual atoms. Helium makes up most of the remaining 10 percent. That leaves only about 1 percent for all the rest. Most things we are in contact with (our Earthly surroundings) consist of this other 1 percent. The reason is straightforward. Most matter in the universe is found either in massive stars (energetic hydrogen/helium balls) or in the huge intergalactic clouds of hydrogen (mainly in the form of plasma).

In fact, we could say that all elements (all physical things) are really just hydrogen. All other elements are formed in the thermonuclear reactions of stars. (Hydrogen fusing to helium and then, in supernovas, to more complex forms.) Therefore, we will simplify the universe and view it as containing solely hydrogen atoms.

The hydrogen atom consists of a nuclear core—a proton—surrounded by an electron cloud. The cloud is much larger than the proton core. (If the proton were a grain of sand placed at the 50-yard line of a football field, the electron cloud would extend to the goalposts.) But we also know that the proton core is many times as massive as its surrounding electron cloud (about 2,000 times as great).

Protons

There are, supposedly, about 10^{80} (100 million, trillion, trillion, trillion, trillion, trillion, trillion) protons in the universe. Most, as already noted, are found as the centers of hydrogen atoms. What, then, is a proton?

All protons are the same. They are in the nuclei of all atoms and their distinct number is what differentiates one element from another. Modern theory considers the proton to be a complex assembly of more basic particles—*quarks.* These quarks are in constant motion but kept together by the *strong nuclear force* that, in turn, acts via other particles—*gluons.* This is all part of the *Standard Model,* a "hodgepodge" of about 20 different variables—today's accepted wisdom. According to the Standard Model, the quarks' mass makes up only about 2 percent of the total mass of the proton; the rest comes from the kinetic energy of the gluons. (As Einstein showed, energy and mass are interchangeable, $E=mc^2$.)

—

The proton has a positive electrical charge. However, in stable atoms the much less massive electron has an equal, negative charge. The neutron, which is found in all elements other than hydrogen, is, as its name implies, electrically neutral. The Standard Model considers it similar, in many ways, to the proton; both contain quarks and gluons. Finally, in the Standard Model, the strong force, keeping protons and neutrons intact, also holds them in place within the nucleus of atoms.

Although this is the current concept, given that the universe is a hypersphere, a different one is now proposed.

Chapter 7

Hypersphere

ALL SPHERES, OF ANY DIMENSION, CAN BE described (mathematically) as simple circles.* The hypersphere, our universe, is a "round" surface composed of the three dimensions of our understandable world covering

---

In mathematics, a round object of any dimension is called a "ball," its surface a "sphere." The surface is always one dimension less than what it covers. All spheres starting with a common circle (1-sphere) and including our universe (hypersphere or 3-sphere), can mathematically be expressed as the product of the circumference of a circle ($2\pi r$) and the interior of a sphere two dimensions less. Thus, our highly complex 3-sphere universe is the product of a common circle's circumference and the interior of an object two dimensions smaller than itself ($2\pi r \times \pi r^2$ or $2\pi^2 r^3$). The actual calculations are not important; what they signify is that all spheres, including our universe, can, from a mathematical standpoint, be considered simple circles.

the unknown fourth dimension of time.

Each instant, a new three-dimensional world appears. It lasts the slightest amount of time—one Planck moment, or 5.4×10^{-44} seconds. (There are about 20 million, trillion, trillion, trillion such moments each second.) It, then, disappears to be "immediately" followed by a new one. This has been ongoing "forever." Since we understand the fourth dimension to be within the smallest possible segments of the three-dimensional universe, we find it inside the Planck volumes of our world. Thus, the Planck volumes contain all of time (all the prior worlds that make up the fourth dimension).

Hidden Energy

The fourth dimension is also the cause of the immense, hidden *zero-point energy* of the cosmos.*

Zero-point energy is believed to be the energy present at absolute zero, when there is no motion or measurable energy. It is estimated to be enormous—10^{120} times that of observable energy. Since the universe is 10^{26} meters (13.8 billion lightyears from any point to the beginning of time), and a Planck volume is 10^{-35} meters, approximately 10^{60} Planck volumes, if lined up side-to-side, would equal its radius. (Although the amount is closer to 10^{61}, I am rounding down to 10^{60} for ease of discussion.) Therefore, 10^{60} cubed (10^{180}) would more or less fill its volume. Since each Planck volume contains 10^{20} times the mass of a proton, and since mass equals energy ($E=mc^2$), then the energy content of our universe is $10^{180} \times 10^{20}$ or 10^{200} times that of a single proton—

This energy can be seen in the tremendous, potential mass of each Planck volume. However, this energy is, for the most part, unfelt, as there is no consistent acceleration of empty space. (Force, as Newton showed, is only present when there is acceleration.) We know that this energy exists, as it manifests itself with the almost instantaneous appearance and disappearance of exotic particles. Thus, empty space is far from "empty"; it only seems that way as the coming and going of particles is not readily discernable to us.

Surfaces

Since our universe is a hypersphere, and since it presents moment-to-moment as a globe (or normal three-dimensional sphere), it should, like any other ordinary sphere, have a two-dimensional surface. Nonetheless, we cannot find an edge to our world: All we sense is it going back in time to the "beginning" or Big Bang. The proton, however, small as it is, serves the function of such a surface.

There are estimated to be 10^{80} protons in our world; and the proton is, by far, the most significant source of energy in the observable universe. Consequently, the untapped or unknown energy—zero-point energy —is $10^{200}/10^{80}$ or 10^{120} times what is known. (Although there is an almost equal number of neutrons, with similar mass, doubling a number as great as 10^{80} makes no appreciable difference in these calculations.)

The proton is 10^{-15} meters in diameter. About 10^{41} protons (rounded down, for ease of discussion, to 10^{40}) equals the radius of the world, and that number squared (10^{80}) would more or less make up its surface. (That same number also happens to equal the total number of protons in the world.) Protons (along with a similar number of neutrons) make up most of the "known" energy of the universe (energy equals mass, $E=mc^2$).

We know that, although the Planck volumes have much greater mass than either protons or neutrons, their presence is not discernable. There is *no net acceleration* of Planck volumes. The mass found in protons and neutrons, however, is felt. It makes up all perceptible matter. The basis, therefore, for tangible mass (and energy) is found in the acceleration (centripetal rotation) of these nucleons. It is what maintains the physicality of objects that comprise our world; it gives the world its surface.

If we were to take every proton from every atom, leaving only empty space, we could construct a hollow sphere the size of our universe. Its entire surface would be protons, its interior completely empty. Our world, however, is fourth-dimensional; consequently, although at any moment it appears three-dimensional, it has no external boundary. Nevertheless, it has a surface, the proton; but that surface is its center.

Chapter 8

Spheres Within Spheres

J UST AS FLAT, TWO-DIMENSIONAL BEINGS WOULD visualize a three-dimensional globe as circles within circles,
we three-dimensional beings visualize a hypersphere as
spheres within spheres. Just as a globe must have the same
strength of compression on its surface, no matter how small
an area, so too does our universe, the hypersphere.

Beachball World

Since this concept is hard to understand, a simpler model
will be used—a universe of only two dimensions, the surface
of a globe. To its inhabitants it would appear as a never-ending
flat plane. Although believed to be flat, it is really the curved

surface of a three-dimensional ball—a "beachball," for example. Its inhabitants can only look out to 90^0 from any point on its surface. Light from beyond would disappear. Objects almost 90^0 away would be tremendously stretched, presenting as a microwave background, just as is seen in our world.

This beachball "world" contains energy in the form of compressed air. That air is equivalent to our fourth dimension—time. The surface strength must be equal at all points; if any site is not strong enough, it would leak and the entire world could collapse. The smallest area, therefore, exerts the same force as does the entirety. Please refer to the following drawing:

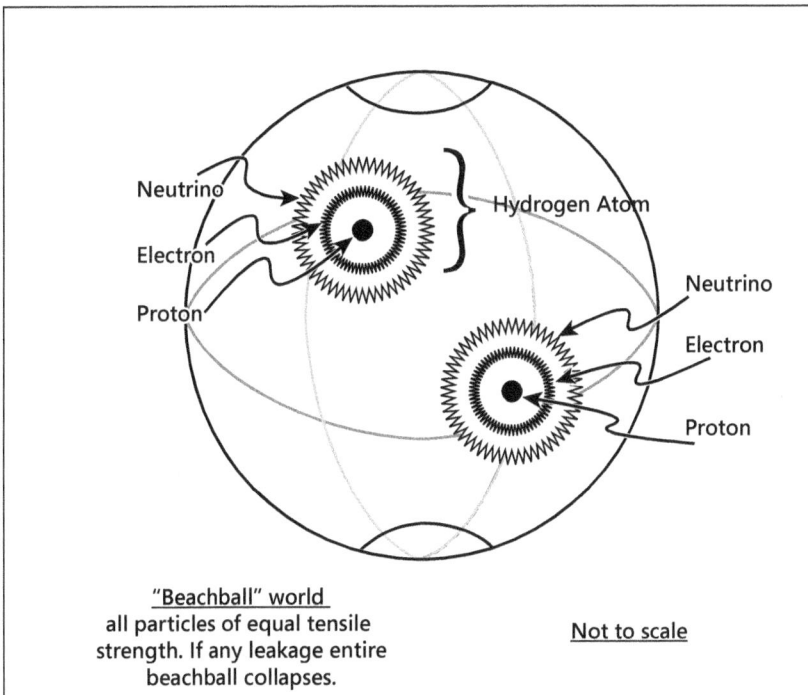

"Beachball" world
all particles of equal tensile
strength. If any leakage entire
beachball collapses.

Not to scale

Centripetal Force

We sense this in an exactly like manner. The ability of a surface to hold within itself its highly pressurized or energetic core is seen as centripetal force (the force pulling toward a center). Mathematically, it is:

$F = mv^2/r$,

where F is centripetal force, m is mass, v is the velocity of rotation of the surface (squared), and r is the radius. From this formula we see that the smaller the radius (r), the greater the force. Since the smallest radius is the Planck length, Planck volumes should, if accelerated, contain the most force.

However, any small, stable particle would be enclosed by vast centripetal forces; thus, the compressive (centripetal) force on the surface of a proton is tremendous—it is greater than even the enormous gravitational force of a neutron star. It (the strong nuclear force) holds a proton together and maintains the integrity of all nuclei.

Neutrons—the "Other" Side

Nuclei, other than the hydrogen atom, also contain neutrons. A nucleus with two distinct nucleons is best understood using the concept of a surface. All surfaces (this page, for example) must have two sides.

Although the proton *is* the surface of "our" world, since all surfaces have two sides, its manifestation on the "other" side is the neutron. The neutron only exists within the confines of a nucleus. When freed, as in beta decay, it rapidly (within 15 minutes) morphs into a proton, an electron, and an antineutrino. It is the "other" side of the two-dimensional edge of our three-dimensional world.

Just as a proton is surrounded by an electron cloud, the neutron has a similar cloud. However, it is on the "other" side of the world's surface. We cannot visualize this other side, as it faces the inscrutable fourth dimension. All we can understand is our side—our three dimensions. So, we perceive the other side as within what already exists. Since all that physically exists are surfaces, already understood as protons, the other side forms "inside-out" protons (antiprotons) surrounded by "inside-out" electron (antielectron) clouds. The other side, as it is sensed inside-out, has an opposite charge (the "other" side of the "real" three-dimensional charge.)*

---

*Another way to understand "charge" is with the yin and yang of force and counterforce. All centripetal force is defined as positive, centrifugal counterforce as negative. Hence, the proton, pulling all things to itself, is positive, and the electron, pushing away to maintain its existence, is negative. If things are reversed and the outside becomes the inside (that is, the electron becomes the center and the proton the periphery), then the pull is still toward the center, or electron, and it now becomes positive (antielectron or positron), and the push, to maintain existence, is in the periphery, or proton, and it is now deemed negative (antiproton).

Antimatter, then, is matter on the "other" side of our world's surface. However, the antielectron cloud, as it is farther away from that surface into the fourth dimension, is found farther "inside" that surface (within the antiproton); and the two, along with an antineutrino, make up the neutron.

The neutron is a hydrogen atom, but visualized inside-out. The "inside" surface of its nucleus (proton) becomes its "outside"; and its far-distant electron cloud (and neutrino) are found deep within that core, all reversed (or inside-out) with opposite charge.

Proton's Size

We can also see why a proton is its distinct size—10^{-15} meters. It is because its potential energy (zero-point energy) must equal the total, observable energy of the world (the mass of 10^{80} protons).

We know that the Planck volume is 10^{-35} meters. Therefore, 10^{20} Planck volumes equals the radius of the proton, and approximately that number cubed (10^{60}) would equal its volume. Since each Planck volume has 10^{20} times the mass of a proton, we have the equivalent, potential mass/energy of 10^{60} x 10^{20} or 10^{80} protons within each proton. This is the same mass/energy as that of the entire tangible world. (*see footnote on the next page.*)

Therefore, if the proton were smaller (or larger) than 10^{-15} meters, it could not contain the appropriate mass/energy. It would be unstable and would not exist. It becomes the essential, smallest, stable particle in the universe. It is the basis of the hypersphere, the edge of our three dimensions, and it constitutes all the tangible substance of the world.

The mass/energy of the observable world is essentially the total mass of all the protons and neutrons. Although there are approximately an equal number of neutrons to protons, doubling a number as large as 10^{80} makes no appreciable difference in the overall effect.

Chapter 9

Classical Theories of Gravity

NEWTON SHOWED THAT THERE IS an intrinsic force of attraction, in all matter, that decreases exponentially with the square of distance.* He knew "how" it acted but could not explain "why" it did so without direct contact.

Einstein considered that gravity was due to acceleration and the bending of space. (The 1919 solar eclipse proved his concepts to be correct.)

If something is two times farther away from a surface, it is attracted with one-quarter of the original force, and if three times the distance, that pull is only one-ninth as potent. The reason is that, although the force of attraction is always the same, it is diffused over the area of a sphere. Thus, the force is the inverse of the radius, or distance, squared. Consequently, the force, although always constant, continually changes with distance. Therefore, any object moving at a steady velocity is met with an ever-greater force; it is accelerated as it approaches.

—

However, Tesla never accepted Einstein's ideas. He felt that the "nothingness" of empty space could not be acted upon (or bent) by "something."

Recent Theoretical Studies

Nevertheless, that "something" causing gravitational change has recently been described. Work out of California and Texas by Alfonso Rueda, Bernhard Haisch, and H. E. Puthoff has shown that Newtown's basic ideas were correct. Newton just was not aware of the particulate nature (Planck volumes) of empty space. Planck volumes have tremendous mass (10^{20}, or 100 million, trillion times as great as protons); but, since they are not accelerating, they are not felt. They make up our three-dimensional world. They are within even the smallest tangible objects. They are the basis of the electron cloud and the neutrino. They are fundamental to gravity and to inertia.

Gravity Is Acceleration

Planck volumes can only be "felt" if accelerating. The attraction of the proton, diffused beyond the electron cloud (and neutrino), pulls on the Planck volumes of space. It is an accelerating force (it increases exponentially). Although exceedingly weak, it is readily felt near massive bodies. The Earth, for example, has about 10^{50} (or 100 trillion, trillion,

trillion, trillion) protons. Objects, thus, are pulled to the Earth's surface ever more quickly the closer they approach.

The basis of this phenomenon is Newton's Second Law: *Force equals mass times acceleration.* The Planck volumes are the mass, now felt, having been accelerated by the huge volume of protons. Newton was correct; he just did not know that "empty" space was *not* empty. Force applied to space, therefore, causes a counterforce. "Empty" Planck volumes now obtain mass as they are pulled toward the proton source. Therefore, that force, emanating from all protons, but not *fully* neutralized by their electron clouds (and neutrinos), is gravity. *It is the acceleration of empty space.*

Inertia Is Acceleration

Inertia is the same. If an object suddenly accelerates, inertia comes into play. Let us say a subway train abruptly starts: Its strap-hangers are jerked backward by inertia. Newton thought this was an intrinsic quality of mass. He understood the action, he just did not understand its cause.

Given recent theoretical insight, we now know that its basis is Newton's Third Law. The action, or force, is the sudden acceleration into "empty" space. But we know space is far from empty. It consists of massive, but quiescent Planck volumes. Once they react, as they must, there arises a counterforce exactly equal to that of the original action. When a suddenly accelerating car pushes you back into the seat, Planck volumes are what react and counter it. Their mass, now awakened, is felt.

Gravity and inertia are therefore two sides of the same coin. We feel gravity as space is accelerated through us, pulled by the constant force of vast collections of protons. We sense inertia as space, reacting to our motion, accelerates back upon us. They both are the acceleration of far-from-"empty" space, which is filled with vast quantities of zero-point energy but sensed only when accelerated.

Chapter 10

Recapitulation

THE UNIVERSE IS A HYPERSPHERE. Its three dimensions are simply the covering of an exotic fourth dimension—time. There is a real distance, infinite in length, within our world's smallest segments. It can only be considered *imaginary*; mathematically it is represented as "*i*", the square root of minus one.

We know that the world *physically* exists. It is the counterforce to the 10^{80} protons that establish all tangible surfaces. Each proton's pull causes an opposing force or push—an electron cloud (and associated neutrino). These almost neutralize the proton's attraction, but there remains a minimal leakage of force. This force is gravity.

Gravity causes massive clouds of protons to aggregate into stars and galaxies. Over eons, these cosmic bodies breakdown and reaccumulate as vast clouds of plasma. This is a continual cycle and it has gone on "forever."

—

Picturing Our World

So how can we understand our world? The best way is to use a simpler example—a two-dimensional model. In this model, each proton—the smallest tangible thing—will be considered sentient. Each "sees" and attempts to "understand" its surroundings. Each finds itself as the center of its world, the farther out the more redshifted objects become. Finally, all things fade into the cosmic microwave background; then there is nothing.

What each of these protons sees is the two-dimensional surface of a globe, considered flat, but actually bent toward the third dimension. This view always disappears at 90^0. Let us draw our globe and place a proton at the North Pole:

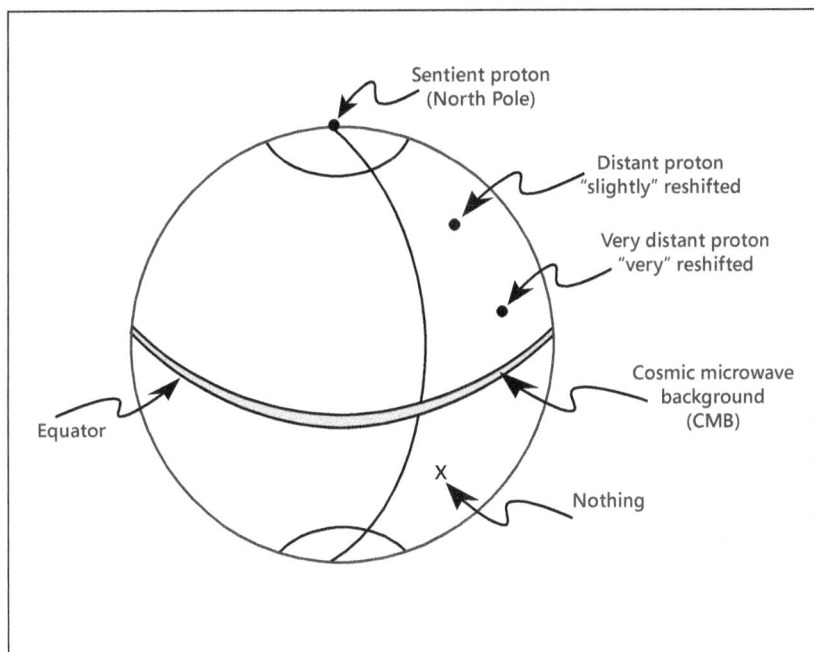

Sentient proton (North Pole)

Distant proton "slightly" reshifted

Very distant proton "very" reshifted

Cosmic microwave background (CMB)

Equator

X

Nothing

Every other proton, no matter how near or distant, is redshifted when observed: the farther away, the greater the distortion. However, every proton sees to 90^0 from its own spot. Therefore, each proton becomes the center of a circle the circumference of which is a hemisphere.

Hence, if we were to include all protons on that surface, each would sense its world end in these circles. In totality, all circles, if drawn, would represent a sphere in three dimensions (globe).

Each proton appears to all others to be from the past.* At each moment, from any proton, the world is a flat circle. *All* other protons, even adjoining ones in the same nucleus, are redshifted; only, the farther away, the greater the distortion. Finally, there is CMB, then nothing.

Since it is a globe, a three-dimensional construct, we (three-dimensional beings) can readily visualize it. As inhabitants of a "three-dimensional" world, the plight of the sentient protons, attempting to comprehend their world, is fully understandable. Furthermore, since all points are on a sphere, they

---

*In our "real," three-dimensional world, everything around us is only sensed by the electromagnetic or light waves emanating from it; all that we "see" is from the past. If some object is 1 meter away, it takes light travelling at 300 million meters per second (3 x 10^8 m/s), one 300 millionth of a second to reach us. So, what we see has occurred in the past, 3 billionths of a second ago. Thus, all things about us are from the past.

are also, by definition, equally distant from one central point. However, this pivotal site, the center of the sphere, is only found in the higher or third dimension, a dimension that "does not exist." Finally, each proton feels a pull toward itself, leading to ever-greater accumulations of other protons (galaxies, stars, and planets). This force is understood as gravity.

Movement In Time

All protons on this two-dimensional model attract all others. Since it is a sphere, all are equidistant to its center. Moreover, being self-aware, all consider themselves to be that center. Every proton is separated from every other one, by time. Thus, all protons become that one central proton (in a higher dimension), but independent and distinct, moving about their surface world, in time.

All protons, then, are one. Einstein's "spooky action at a distance," the whole basis of quantum theory, can now be explained as there being but one object. It, however, is in a higher dimension.

Let us view this now in our "real" three-dimensional world. From every proton the world is sensed as a globe, extending out 13.8 billion lightyears. The farther out, the greater the redshift. Finally, there is CMB, then nothing.

Big Bang proponents believe that all things began at a moment, 13.8 billion years ago. But the James Webb Space Telescope will show that, no matter how far away or back in time we look, things are essentially the same. It will show an absence of early, immature galaxies even at the very edges of time. It will show *déjà vu, all over again.*

Our world, from any proton's perspective, is a globe (a moment in time) that encompasses all other protons. Just as in the lower-dimensional model, all protons being equally distant to an edge signifies that they are one, but now in a higher or fourth dimension. Separation of protons is really due to *motion in time.* This movement takes place in the fourth dimension *between* each Planck moment.

Each proton attracts all other protons. This leads to massive accumulations (galaxies, stars, and planets). Force is seen, on a cosmic scale, as gravity. This force is also felt on a submicroscopic level. All forces (gravity, weak, electromagnetic, and strong) are the same. All are aspects of the centripetal pull toward a center. *(See footnote on next page.)* Without this pull surfaces could not form and nothing tangible could exist.

The true center of our universe is, therefore, in the higher or fourth dimension. It encompasses all protons, for all are really one. The universe, then, can be explained as an emanation from a single source. It cannot be understood in three-dimensional or normal terms.

Perfect Cosmological Principle

What we have shown, then, is the *Perfect Cosmological Principle.* There is no single center to the universe. *All* points

—

are equidistant to its edge—the beginning of time. But that edge is not a Big Bang; there was no cataclysmic beginning. It is merely 90^0 on a mysterious hypersphere.

Thus, given a force of attraction (the centripetal force of each proton), electrons, neutrinos, and the remainder of the universe are its counterforce. Gravity is the residue of the force emanating from each proton felt at the very edge of the universe. It causes an accumulation of protons leading to stars and galaxies. It is why we exist. The difficulty in visualizing its action is due the universe's infinite, hidden fourth dimension. We can only grasp at an understanding; we can never fully comprehend it.

Centripetal force measures pull toward a center. It is what maintains the rotation or orbit about a central point. Its formula is mv^2/r, where m equals the mass of a Planck volume (the smallest possible particle of space), v equals its velocity of rotation squared (in the case of the fundamental forces, we can assume it to be c—the speed of light), and r is the radius of orbit. The greater the radius, the greater the circumference or distance of that orbit, and the longer a Planck volume, moving at a constant velocity, will take for completion. If force is measured as impacts per second (at some set location in an orbit) then the greater the distance traveled (to complete that orbit), the fewer impacts per second and the weaker the force. Consequently, since gravity, and the weak, electromagnetic, and strong forces are all caused by centripetal pull, their strengths are proportionally lessened by their distances from a center. Hence, the strong force is 10^{41} times greater than gravity, because the universe is 10^{41} times larger than a proton. It is the same force, but diminished by a great distance.

Chapter 11

Summary

G RAVITY, EMANATING FROM every proton, is exceptionally weak. Nevertheless, it is felt if sufficient protons are present. It extends to the very reaches of the universe. It is a constant pull back to each and every proton and would, over time, cause the universe to collapse. However, the universe maintains its spherical shape, as there exists an equal counterforce—the highly energetic plasma clouds of hydrogen, scattered throughout intergalactic space.

Just as stars maintain their shapes through thermonuclear reactions, so too does the universe with energized plasma. The overall shape, however, is that of a hypersphere. Every proton is its center. The world extends outwardly to the limits of gravitational attraction from every point. All things are equally distant from the edge of time.

What, then, will the James Webb Space Telescope show us

as we look to the very boundary of our three-dimensional world?

We will find that, wherever we search, the universe is the same. There was no explosive start—no Big Bang. The redshift is simply the acknowledgement of a higher-dimensional curve. The cosmic microwave background is merely the farthest edge of that curve, just before it disappears. Inflation is an *ad hoc* addition with no basis; it is not needed. Dark energy is a chimera, a mythical beast, created for a nonexistent Big Bang; objects in distant space *are* where they should be.

The universe follows the *Perfect Cosmological Principle*. It is the same everywhere and has been for all time. There was no beginning. There will be no end. It is truly *déjà vu, all over again*. The JWST will finally rid us of Big Bang foolishness; science will end a 100-year misadventure. Gravity and inertia will be fully understood, and we will advance into the 21st century with all the wonders of a true, space-faring culture.

Appendix A:
Dark Energy

As already noted, all surfaces, all spheres, are circles ($2\pi r$) multiplied by interiors two dimensions less than what is covered. Hence, the universe as a 3-sphere ($2\pi^2 r^3$) is really just a circle, and a tangent to it can be reduced to our simple example of a one-dimensional straight line.

Thus, if we take the visible universe of 13.8 billion light-years to be a quarter of a circle (0^0-90^0) and divide it equally, we get constant distances two-dimensionally that continually elongate one-dimensionally. Let us, therefore, divide ¼ of a circle into 5 equal portions; each is 18^0 ($90^0/5$) and equals 2.76 billion light-years (13.8/5). However, although these portions are always the same, the tangents increase as we go farther out. The following illustrates this concept:

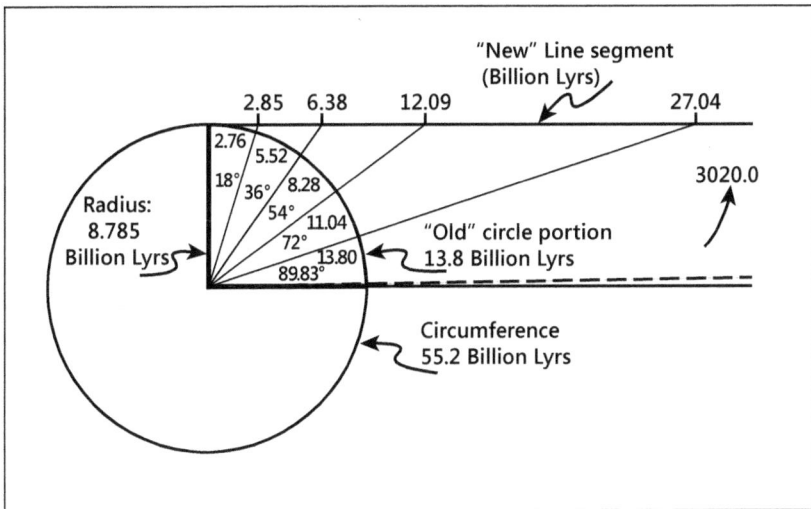

A *tangent* is defined as "opposite/adjacent in a right triangle." In our example, "opposite" is the new length on the straight line, and "adjacent" is the radius of the circle comprising our universe. To find the radius, all we need is the total circumference of our circle. Since 13.8 billion light-years is ¼, the total is 13.8 x 4, or 55.2 billion light-years. Now, $2\pi r$ = circumference, or, if we cross-multiply, r = circumference/2π; since π = 3.14159, 2π = 6.28318; therefore, radius, or r = circumference / 6.28318; or, r = 55.2/6.28318; or, r =

8.785 billion light-years.

Thus, because tangent = opposite/adjacent, tangent = straight line/radius, or (cross-multiply):

Tangent x radius = straight line.

Hence, all we do is multiply the tangent of each angle by 8.785 billion light-years.

ANGLE	TANGENT		RADIUS		LINE SEGMENT	CIRCLE PORTION
18^0	0.32492	x	8.875	=	2.85	2.76
36^0	0.72654	x	8.875	=	6.38	5.52
54^0	1.37368	x	8.875	=	12.09	8.28
72^0	3.07768	x	8.875	=	27.04	11.04
89.83^0	343.774	x	8.875	=	3020	almost 13.80

We now have the straight-line segments, and to get the z values we use the formula:

Z = new – old / old

Where "new" is the line segment, and "old" is the circle portion (each 2.76 billion light-years). We now have a z for every tangent, thus a z that matches each distance for the 4D curve/tangent column.

The distances of the Hubble/expansion column are based on the following formula:

Z = square root (1 + v/c / 1 − v/c) − 1.

To get the velocity (v) using Hubble's constant of expansion (the current best estimate is 67.8 km/sec for every 3.26 million lightyears [megaparsec]), we take a distance (let us say 1.5 billion lightyears) and divide it by 3.26 million to get the number of times its velocity has increased, then multiply by 67.8 k/s for each increase. Thus, in the example, 1.5 billion light-years is approximately 460 (1.5 billion/3.26 million) times 67.8 k/s, or about 30,000 k/s. We then put this velocity into the above formula:

Z = sq rt (1 + 30,000/300,000 / 1 − 30,000/300/000) -1; or, z = sq rt (1.1/0.9) − 1; or, z = sq rt (1.22) − 1; or, z = 1.1 − 1; or, z = 0.1.

Thus, for any distance we can get a z value according to Hubble's law of increasing velocity with distance. When we compare the two columns, we find distance to be greater for the 4D curve/tangent than for Hubble/expansion at every z until about 2.0.

Hence the universe is really bigger, because of a fourth-dimensional curve, than thought due to simple expansion. Dark energy, therefore, has no rationale. It is not required, as there has been no initial or additional expansion. It is an illusion.

www.ingramcontent.com/pod-product-compliance
Lightning Source LLC
Chambersburg PA
CBHW042118190326
41519CB00030B/7543